F

FOR YOUR LIFE!

FIGHT FOR YOUR LIFE!
The Secrets of Street Fighting

Dr. Ted Gambordella

Barricade Books • Fort Lee, N.J.

Published by Barricade Books Inc.
1530 Palisade Avenue, Fort Lee, NJ 07024
Distributed by Publishers Group West
4065 Hollis, Emeryville, CA 94608

Copyright © 1982 Ted Gambordella

All rights reserved.

No part of this book may be reproduced, stored in a retrieval system,
or transmitted in any form, by any means, including mechanical,
electronic, photocopying, recording, or otherwise,
without the prior written permission of the publisher, except by
a reviewer who wishes to quote brief passages in connection
with a review written for inclusion in a magazine,
newspaper, or broadcast.

Printed in the United States of America.

Library of Congress Cataloging-in-Publication Data

Gambordella, Theodore L.
    Fight for your life!: the secrets of street fighting/
  Ted Gambordella
Originally published: Boulder, CO: Paladin Press, 1982
ISBN 0-942637-65-8 : $9.95
    1. Self-defense. I. Title
[613.6'6—dc20   92-14947
                                         CIP

To Zlatko Chiddix and Barry Guimbellot for their help and assistance. To Donna McCorry for herself, and to the martial artists of Texas, the greatest in the world.

Contents

Foreword	ix
Introduction	1
1. Fighting Techniques	3
2. Escapes and Counters	19
3. Close Fighting and Ground Fighting	29
4. Knife and Club Defenses	45
5. Defenses for a Woman	73
6. Multiple Attacks	83
7. Tips	93

Foreword

Fight for Your Life: The Secrets of Street Fighting is part of a series in which I show the serious student of the martial arts and the teachers techniques that can be of practical value in the street as well as in the dojo. In order to fully appreciate the material, you should obtain my book, *The 100 Deadliest Karate Moves*, for reference as to the vital points we will be discussing in this book.

I did not write this book to teach readers how to beat up other people, or to train would-be muggers. The techniques found here are defensive, and although they are quite deadly, they will not work quite as well in offensive, attack situations.

The techniques found in this book require a basic knowledge of karate or jiu-jitsu, but they should be familiar to all students of the arts who have a minumum of six months of study and training. I do not present flashy techniques or superfluous finishing touches, unlike other books where you see the defender kick the attacker twenty times, strike his eyes, tear out his throat, flip him, and then crush his chest. My secret techniques that you are about to discover will produce sufficient damage to stop the attack and to make certain that it cannot continue.

I do not advocate killing anyone for any reason unless you are certain that you are about to die, or unless the person is in the act of killing another person.

In closing, allow me to say that as a Christian martial artist, I study the arts and write about them as a teacher and artist. Note that I am not a proponent of violence as a way of life. But if you practice these techniques in good faith, I am confident that they will enable you to win that crucial street fight engagement, where you must win or die.

Introduction

Fight for Your Life: The Secrets of Street Fighting concerns just what the title implies: learning the secret art of fighting in the street, winning at all costs and under any circumstances. One of the biggest criticisms of martial arts in America is that it does not work in the street, and I am sure that you have all heard of a supposed Black Belt who got the hell beat out of him by some football player or streetwise hoodlum. I am also sure that you have probably seen many a young student of karate doing his stuff, and you have thought, "I could kill him if I wanted to. Who does he think he's fooling?"

Well let me explain to you why this happens and what can be done to prevent it. What you must first understand is that American karate is not Japanese karate, and Japanese karate is nothing like the American art. Let me explain.

In Japan, only the best athletes are allowed to train in a true karate school. In America, on the other hand, the best athletes play football. In Japan, they train six hours a day, seven days a week, for up to five years before they are given the rank of Black Belt. In America, you can get a Black Belt at most schools for training two to three years, three times a week, for two hours a day. Finally, in America, fights in the street are commonplace, and the people who usually start the fights are trained and skilled in street fighting. These are usually snakelike, vulgar individuals who will use any punch or kick necessary to win. In Japan, only a dishonorable person would start a fight in the street.

The honorable man would not ever even consider losing face and fighting in the gutters over a drink or a woman.

Thus, we can see that in America, dirty-type street fighters are common. The karate-ists who they fight are not the best athletes around. But in Japan, the best athletes *are* the karate-ists, and there are essentially no street fighters who do it daily and just for fun (of course, there are gangsters, but we are talking about barroom fighters). So when the street fight does happen in America, the street fighter usually wins. The simple reason for this result is that he has been trained in the art of dirty fighting and has no Christian or social morals to constrain him from killing or crippling you. Also, he usually has the great advantage of starting the fight.

This book can change all that for you. Now you *can* win in the streets, because the secret techniques that I am showing are effective, simple, and require no great physical strength; and while they will definitely stop the attacker, they are not meant to kill him.

If you will practice a few simple kicks, stay flexible, and train daily until the moves presented herein are second nature, then should the time ever come when some poor fool attacks you in the street, the surprise will be on him. He will find himself waking up in the hospital, or worse.

1. Fighting Techniques

You are attacked with a hard right by a punk.

You react by blocking with your left, and then at the same time coming up with a front snap into his groin.

Finish him off with an elbow to his throat.

You are attacked by a boxer type, who leads with a left jab.

You block with your left, and at the same time snap up a side kick, breaking his ribs.

Finish him off with a knee to his jaw and face.

Again you face a boxer type. This time as he fakes at you, snap a side kick to his knee.

Now step up and grab the back of his head, and smash his face with a hard elbow.

This time as the attacker begins to get into his defense stance, you thrust a roundhouse into his solar plexus, stunning him. Then move in and smash his stomach again with a left.

Fighting Techniques

Continue to counter with a hard right to his temple, and then knock him out with a hard left hook to his jaw.

As he gets into his stance, you snap up a front roundhouse to the side of his face. Then you quickly step in and grab his neck in a hip-throw position.

Fighting Techniques 11

Now you bend him backwards, smash his back and neck against your knee, and then reach down and grab his face and tear out his eyes.

You face him off and lead with a quick stinging left to the side of his face. This blinds and stuns him enough to allow you to move in with a foot sweep to the back of his knee.

Knock him to the ground, where you finish him with a face stomp.

As he throws a jab at you, snap up a front kick into his ribs.

Now move up and quickly use your knee into his groin. Then finish him off with a head smash into his nose, knocking him out.

The attacker throws a kick at your side, which you block with a left lower circular block. Then counter his kick with your own into his groin.

Fighting Techniques 15

Now you step up and throw a hard shuto into the back of his neck. Then finish him off with a knee smash into the throat.

He attacks with a right fake, and you react by snapping a hard front kick into his throat.

Fighting Techniques 17

You then turn and smash a heel kick into the side of his face. As he falls backwards, finish him with a knee break.

2. Escapes and Counters

You are caught in a headlock. React by smashing your left into his groin. At the same time, your right comes up and over his head, grabbing his hair.

Pull his head back to expose his body for your finishing punch, which is a smash into his throat, or you can hit his groin.

You are attacked from behind and grabbed in a full nelson. Bring your arms up to your head to relieve the pressure. Now come down with a heel stomp to his instep.

This will cause him to release his hold. Now step out and smash an elbow into his stomach. Finish him with an elbow to his jaw, breaking it.

You are caught in a front grab around the waist. Before he can knock you to the ground, come up and smash your hands to the back of his neck.

Finish him off with a knee to his chest.

You are grabbed in a front bear hug. Raise your arms up very high to get power, and smash his eardrums.

Now come up and smash his nose with a backfist, and then forcefully apply an elbow to his jaw. Finish him with a front kick to the groin.

Escapes and Counters

A lady is grabbed from behind in a bear hug, but her arms remain free. She swings an elbow to the side of the attacker's face, and then the other elbow to the other side.

Now she leans over, grabs his leg, and throws him to the ground by pulling his leg up. She finishes him with a heel kick to the groin.

Escapes and Counters

The attacker reaches up and grabs a lady's hair, preparing to strike her.

The woman reacts with a rising block.

She counters with a snap to the groin.

3. Close Fighting and Ground Fighting

Your attacker reaches at you to grab you.

You duck down and grab his left arm as it crosses your body.

Now come up and break his elbow with your forearm. Finish him off with a knee to his face.

As a street fighter reaches at you, you block his left with a rising right block. Then come up and smash your knee into his face.

Now step around behind him to grab his face and claw his eyes. Then slide up to his neck and choke him into unconsciousness.

As the attacker reaches at you, you duck down and push his left across your body. At the same time, you step around behind him and begin to lock him up.

Close Fighting and Ground Fighting

Now lock him up by reaching under his left arm, and begin to choke him. Finish your move by breaking his back and his neck.

Your attacker runs at you and rams his head into your stomach. React quickly by smashing his neck with your hands. Then lock his neck in your arms.

Close Fighting and Ground Fighting

Now pull up very hard and snap his neck. Then finish him off with a knee to his face.

You are knocked to the ground and are on your back. As your attacker attempts to get on top of you, reach up and grab his hair.

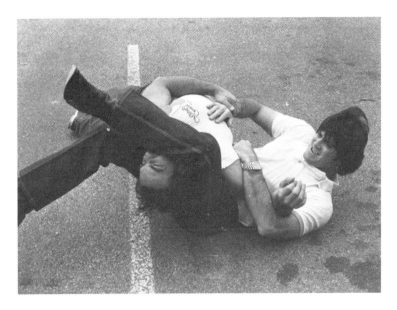

Pull him across your body by his hair, until you can lock his face with your legs and begin to choke him. Then finish him with a hammerfist to his groin.

Your attacker is on top of you and attempts to strike your face. Block with your left, and then come down and smash his groin.

Now reach up and grab his head, pull him down, and smash his face into the ground. Then throw him over yourself by kicking up with your body and pulling with your hands.

Then jump up and smash his face with your heel.

An attacker is on top of you, attempting to strike your face.

Quickly bring up your legs and wrap them around his body.

Now finish him off with a smash into his groin with your backhands.

Close Fighting and Ground Fighting 43

A man is on top of a lady trying to choke her. She quickly reaches up and strikes his eyes with her fingers.

Now she swings her legs up and over his body and throws him backwards, where she finishes him off by smashing his groin with her fists.

4. Knife and Club Defenses

Briefcase Defense

You are attacked by a thug who tries to steal your briefcase as you leave your car. You briefly resist his pull, then quickly come toward him and smash his jaw with an elbow.

Next you smash a knee into his groin, knocking him to the ground, where you finish him off with a smashing heel stomp to his groin.

Knife Defenses

A thug attacks you with a knife, using a straight thrust. You pretend to give up and raise your arm. As he thrusts at you, parry his strike with your left arm.

Now quickly smash a front snap kick into his groin, knocking him to the ground. Take the knife out of his hand and stomp his face or groin, or you may hold him now for the police.

Knife and Club Defenses

The attacker thrusts the knife at you. Quickly close in, blocking the knife before he comes across your body. Now move in and do a hip throw.

After you throw him to the ground, smash his face with a heel kick.

The attacker with a knife makes a direct thrust to your stomach. Step back, and use an X-block to stop the knife. Now roll your hands over, bring them up and over your head, and throw him to the ground. (Close-up of the position of the hands.)

Continue to hold his arm after he is down. Stomp his throat to finish him off.

The attacker holds the knife to your throat. You decide to resist. (Note: I do not recommend that you try to resist this type of knife attack unless you are very proficient in karate, or unless you feel that he will kill you anyway.) You reach up quickly to loosen the knife pressure and bend your neck away from the blade.

Knife and Club Defenses

Now do a shoulder throw and flip him over your shoulder to the ground. Stomp his groin to finish him.

You are attacked with a knife. The attacker is using a side swinging action, trying to cut out your guts. You quickly bring up your left arm to block. As he thrusts at you, parry the blade. Then snap up a crescent kick onto his forearm.

Knife and Club Defenses

Your kick should have him moving forward. You counter now with a shuto to the back of his neck, breaking his neck and knocking him to the ground. Finish him with a stomp to his heart.

The attacker has a knife into your back. Raise your arms up as if you were surrendering. Now quickly swing around to the left and pull your left arm across your body as you move back away from the knife.

Finish him off with a side kick into his chest, crushing his rib cage.

The knife is coming at you from an overhead strike. Raise your left arm up to block, trying to catch his forearm with your wrist area. At the same time, smash a front snap kick into his groin.

Now you lock up his arm (*see* close-up of the locking position), and step behind him and throw him to the ground. There take the knife out of his hand and cut his throat with it, or hold him until help comes.

Knife and Club Defenses

The attacker lunges at you with the knife, using an overhead strike. Moving very quickly, you step back, take off your coat, and wrap it around your arm.

Using the coat for a safety wrap, you block his thrust and snap a kick into his groin.

Knife and Club Defenses

Now come up and finish him off with an eye strike, blinding him until help arrives.

The attacker is threatening you with the knife. Raise your arms as if to give up. When he attacks anyway, quickly step back and grab your belt and take it off.

Using it like a whip, flip the end of the belt into his eyes, stunning him and causing him to lose his vision.

Now quickly wrap the belt around his neck and choke him unconscious.

Knife and Club Defenses 65

The attacker has a knife blade against your throat. Quickly grab the side of his arm and relieve the pressure of the knife against your throat. Now come back with a hard elbow smash into his stomach.

Come back up into his face with a backfist strike into his nose. Now step under and behind his arm, and force the knife blade into his own stomach (this move is easy to do, since he has no strength to prevent you from stabbing him with his own knife in this position).

A man is attempting to remove his knife from its scabbard. You quickly jump forward and grab his arm, preventing him from drawing his knife.

Now come up and smash his eyes with a two-finger strike. Then execute a knee smash into his groin.

Finish him with an elbow to his face.

Club Defenses

You are attacked with a pipe. As the assailant swings at you, duck down and let the pipe go over your head. Before he can come back, smash his groin with a right thrust punch.

Now finish him off with a shuto chop to the back of his neck, breaking his neck.

You are caught from behind in a choke hold by a stick or pipe. Quickly come up and grab the pipe to relieve the choking pressure. Now strike back with your backfist into the attacker's teeth and nose.

Continue to step around and twist your way out. Then strike his groin with a backfist. Finish him off with a knee to the jaw, breaking his jaw.

5. Defenses for a Woman

A lady is attacked with a knife thrust at her chest.

React quickly and bring up your purse to block the knife. At the same time, kick your assailant's groin with a front snap.

Now come across the side of the attacker's face with an elbow smash, knocking him to the ground. Finish him with a stomp to the face.

A lady is attacked with a pipe. As the attacker swings the pipe, quickly bring up your purse to block it.

Finish him off with a front snap to the groin.

A lady is attacked with a front choke. Quickly snap up a front kick to the attacker's groin.

Defenses for a Woman

Now reach up and thrust your thumbs into his eyes. This will force him away, and you can finish him with another kick to the groin.

A lady is grabbed from behind in a bear hug. React quickly, before any pressure can be exerted. Smash your heel into his instep. Now come back with an elbow into his stomach and ribs, knocking him away.

Finish him off with an elbow smash to his jaw, breaking it.

Defenses for a Woman 81

A thug tries to grab your purse. Pull away as if to resist.

Now react with a side kick into his knee.

Finish him with a groin stomp.

6. Multiple Attacks

You are attacked by two men at once. Quickly snap a groin kick to the first man.

Now you smash an elbow to the second assailant. Then finish him with a knee to the face.

Multiple Attacks

You then finish the first man with a back-heel kick to the face.

You are defending yourself against one man with a quick claw to the face, followed by a snap kick to the groin.

You then finish him with a back kick to the rib cage. But his friend decides to help, so you snap a kick to the second attacker's groin, and finish him with an elbow smash.

A lady is attacked by two men. She snaps a front kick into the groin of the man in front of her.

Now she thrusts a back kick into the groin of the man behind.

Then up with an elbow smash to the side of the head of the first man.

She finishes the second man with a heel kick to the heart.

A lady is grabbed from behind and held as a man attacks her from the front. She quickly snaps up a front kick into his groin. As she brings her foot down, she smashes the instep of the man holding her.

Now she comes up with a hard elbow into his jaw, knocking him out. She finishes the man in front with a knee to his face.

7. Tips

A lady is grabbed on the chest by an attacker.

Reach up and claw his eyes.

Finish him with a knee to his groin.

If you find yourself grabbed from behind by a man, smash your elbow up into his face.

If you find a man on top of you, smash your knee up into his groin.

Mace is one of the best and most effective defenses that you can carry.

If a man is trying to slap you around, you can cover by putting your arms up to the sides of your head and blocking his punches.

Then you can counter with a snap kick to his groin.